Climate

by Hilary Maybaum

Table of Contents

6

28

4

What determines climate and what causes it to change?

26

16

YOU WILL SEE DUBAI'S LATEST DEVELOPMENT IN THE DISTANCE...A MAN-MADE GROUP OF PRIVATE, AIR-CONDITIONED ISLANDS. WHEN COMPLETE, THIS MICROCLIMATE WILL RESEMBLE A MAP OF THE WORLD WHEN SEEN FROM ABOVE...

It Wasn't Always Cold Here

▲ The Arctic wolf evolved from a group of dog-like mammals that roamed the land millions of years ago.

You would not have recognized the North Pole 50 million years ago. This period marked the start of the Cenozoic Era of Earth's history. At that time, conditions everywhere were warm and moist. Can you imagine it? The North Pole and surrounding Arctic region were almost tropical! There was no ice or snow to be seen. Back then, lush ferns and small lakes dotted the land. Few mammals existed, and they all weighed less than 10 kilograms (22 pounds). At a balmy temperature of 23°C (73°F), they could have swum comfortably in the Arctic Ocean.

Within a few million years, however, the scene began changing. Temperatures cooled very slowly. Snow replaced rain. The rain collected on mountaintops, forming ponds. Forests of evergreens and broad-leafed trees took the place of ferns. Layers of packed snow slowly compressed into glaciers (GLAY-sherz) on the land, while ice began forming around the polar oceans. Dog-like mammals roamed the Arctic region, finding food wherever they could.

Hubbard Glacier, Alaska

By 20,000 years ago—around the time that humans appeared—Earth was in an ice age. Large ice sheets and massive glaciers blanketed much of North America. Slabs of pack ice floated slowly across the northern oceans. Most of the Arctic Ocean was itself covered by a thick, floating shelf of frozen water. Meanwhile, the regions near the equator (ih-KWAY-ter) remained warm and humid, teeming with life.

An interesting thing began to happen about 7,000 to 14,000 years ago. The ice sheets started to melt. Earth was warming once again. With temperatures on the rise, the glaciers began retreating. Much—but not all—of the ice melted in many regions of North America. However, the polar regions kept their stores of frozen water. Only portions of the ice would melt in summer.

This was not the first ice age in Earth's history, but it may have been the last. Something else is happening now in the Arctic that no one had predicted. Read on to find out more about Earth's changing climate.

Understanding Climate

What is climate, and what factors determine it?

t's a blustery January day in New York's Central Park. Squirrels dig through the snow-covered ground in search of acorns buried months ago. Your friends are eager for you to put on your ice skates and join them on the frozen pond. Good thing you brought your hat and scarf!

On this same day, about 3,575 kilometers (2,240 miles) to the south, the sun shines brightly in a cloudless sky. An outdoor soccer game is taking place in Panama City, Panama. The players wear shorts and T-shirts. They stop to wipe sweat from their brows.

As you can tell, New York City in January is a lot different from Panama City. In one place, the air is frigid—in the other, the air is hot and humid. What makes these two locations so different from each other? The answer is **climate**. Climate is the average set of atmospheric conditions in a region over a long period of time. In other words, climate is the average weather of a region.

Different factors help determine climate. These factors include average air temperature, precipitation, and humidity. Air temperature is the amount of heat energy in the atmosphere. Precipitation is water that falls to the ground as rain, snow, sleet, or hail. Humidity is the amount of water vapor in the air. The climate of any region is based on the effects of these factors over long periods of time.

New York City

In New York City, the average air temperature in January is 0°C (32°F).
In Panama City, where it is warm year-round, the average weather in January is 25°C (77°F).

Panama City

EQUATOR

Comparing Climate and Weather

Climate is the average set of weather conditions in a given region. Average is a math term. An average is the typical value of a set of numbers. To find the average, first add the values in the set. Then divide the total by the number of values. For instance, if you have three numbers, 1, 5, and 6, the average is 4, because:

$$1 + 5 + 6 = 12 \text{ and } 12 \div 3 = 4.$$

The atmospheric patterns of each of Earth's climates are based on years and years of data. These patterns tell us that it will snow in the northeastern United States in winter and become mild in the spring. Weather, on the other hand, is the daily condition of Earth's atmosphere. These conditions change from day to day. Weather can even change from moment to moment.

The Role of the Sun

Without our closest star—the sun—Earth would be a dark and chilly place. Although it is 150 million kilometers (about 93 million miles) away, the sun plays a huge role in Earth's climate. Energy from the sun heats Earth's atmosphere and surface. The sun keeps the planet warm and hospitable to life.

The amount of energy Earth receives from the sun depends on the position of Earth. As Earth spins on its axis, it also orbits the sun. Earth's axis is tilted at an angle of about 23.5°. This tilt makes one pole of Earth lean toward the sun while the other pole is tilted away. This effect, in turn, causes seasons.

The equator is the imaginary line that circles the center of the globe. Areas north of the equator make up Earth's Northern Hemisphere. Areas to the south

✔ CHECKPOINT

VISUALIZE IT

In the diagram below, locate the approximate positions of the two cities discussed on page 6 during the month of January. Describe their locations relative to the sun. Think about what the weather will be like in each place in six months.

form the Southern Hemisphere. In June, the Northern Hemisphere gets the most sunlight. This brings summer weather to North America, Europe, and Asia, and winter to the Southern Hemisphere. In December, when it is winter in the Northern Hemisphere, it is summer in the Southern Hemisphere. The equator always receives intense sunlight. There, it is warm and sunny year-round.

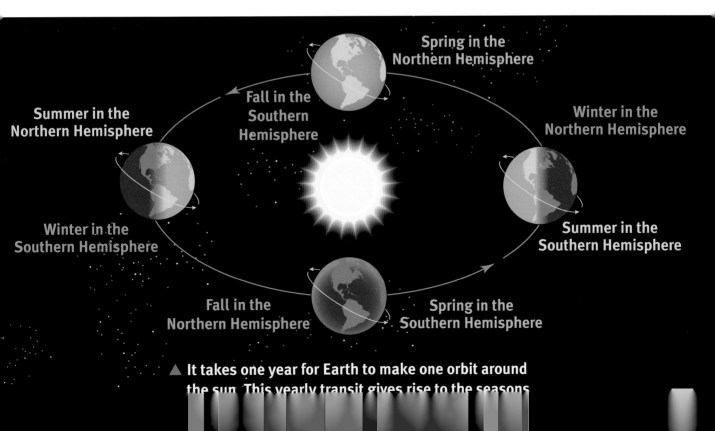

Spring in the Northern Hemisphere

Fall in the Southern Hemisphere

Summer in the Northern Hemisphere

Winter in the Northern Hemisphere

Winter in the Southern Hemisphere

Summer in the Southern Hemisphere

Fall in the Northern Hemisphere

Spring in the Southern Hemisphere

▲ It takes one year for Earth to make one orbit around the sun. This yearly transit gives rise to the seasons.

The Greenhouse Effect

Most of the energy from the sun breaks through Earth's atmosphere. Clouds, particles in the air, land, water, and living things absorb about half of this incoming energy. The rest is reflected back out to space.

When the sun's radiant energy is absorbed, it changes into heat energy. As an object gains more heat energy, its temperature rises. The heat energy then reradiates back out toward space. In doing so, it warms Earth's surface and atmosphere.

Not all of the reradiated heat energy escapes. Certain gases in the atmosphere capture some of it. These gases send the heat energy back toward Earth, causing even more warming. This trapping of heat energy is called the **greenhouse effect**. The greenhouse effect helps control Earth's temperature. The substances responsible for this effect are called greenhouse gases. You will learn more about the greenhouse effect, and its importance to climate, in Chapter 3.

Everyday Science: Watts Up?

The flow of energy can be measured in units called watts. One watt is equal to one joule of energy per second. A 100-watt lightbulb, such as you might find in a ceiling lamp, produces light energy at a rate of 100 joules per second. The energy of the sun is equivalent to nearly 4 sextillion 100-watt bulbs—that's 4,000,000,000,000,000,000,000.

Everyday Science: A Misnamed Process

It was once thought that a greenhouse stays warm in the same way that Earth's atmosphere stays warm. In fact, greenhouses stay warm because cooler air on the outside of the glass windows cannot interact with the warmer air inside. The "greenhouse effect" is therefore not an accurate name.

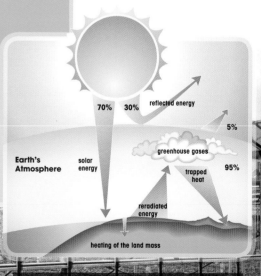

The greenhouse effect keeps Earth warm by trapping reradiated heat energy in the atmosphere.

Factors That Affect Climate

You read on page 6 that climate is determined by air temperature and precipitation. Several factors affect the temperature and precipitation of a region. These are **latitude**, elevation and topography, nearness to water, and winds and ocean currents.

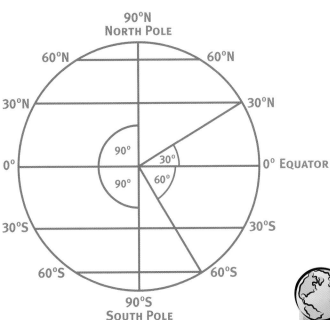

▲ Lines of latitude are drawn parallel to the equator. These lines indicate distance north or south of the equator. Places far from the equator receive less energy from the sun. That is because the same amount of sunlight is spread across a larger area.

Science & Math: The Geometry of Latitude

You are probably familiar with using degrees to measure angles. So how can you measure angles of latitude? First, visualize slicing Earth in half along a great circle. Then draw a line connecting the North Pole and South Pole. This line is perpendicular, or at 90°, to the equator. Next, choose a point at the surface of Earth and draw a line from that location to the center of Earth. The angle between that line and the equator (0°) is the latitude. It is important to use the letter designations N or S on the measurement to show if it is above or below the equator.

Elevation and Topography

Have you ever climbed a mountain? If so, you may have noticed it gets colder as you move higher. The air temperature depends on your elevation. Elevation is height above sea level. For every 100 meters in elevation, the temperature drops by almost 1°C. That's a drop of about 5.5°F for every 1,000 feet you climb.

Mountains and other land features, such as valleys and canyons, form the topography of a region. **Topography** is the way in which elevation rises and dips across Earth's surface. Topography affects the land's temperature and moisture levels.

The island of Kauai is in Hawaii. The island is the site of one of the rainiest places on Earth. More than 1,100 centimeters (about 36 feet) of rain falls on Mount Waialeale (wy-AH-lay-AH-lay) each year. Yet 24 kilometers (15 miles) away, the dunes of Barking Sands get less than 50 centimeters (20 inches) of rain. This effect is known as a **rain shadow**. It is caused by the nearness of the ocean and the height of the mountain.

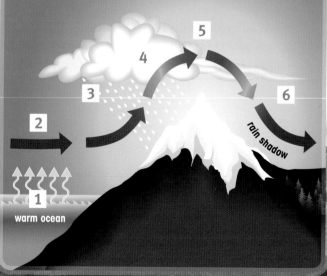

Rain Shadow

1. Seawater evaporates.
2. Moist air is blown toward land and moves up the mountain.
3. As the moist air rises, it cools and forms clouds.
4. Rain falls from the clouds. The air dries.
5. Cool, dry air moves down the other side of the mountain.
6. As the dry air falls, it becomes warmer.

rain shadow

warm ocean

▲ A dry region can exist in the rain shadow of a mountain. The Mojave Desert, in the southwestern United States, is the result of a rain shadow.

Towns and cities along the banks of rivers are cooler and more humid than areas farther inland.

Science and Technology: Remote Sensing

Simply put, "remote sensing" means using a device to observe distant objects. Reading these words is a form of remote sensing. Your eyes are the device, and the words are the objects.

In science, remote sensing involves more complicated instruments—such as spacecraft. These craft also look at faraway objects. This method lets scientists observe, measure, and record data in the atmosphere, oceans, and deep space.

Satellites are another form of remote sensing. They are especially helpful to climate scientists. Some satellites are launched in joint missions between government agencies. For instance, the Tropical Rainfall Measuring Mission is a program of the National Aeronautics and Space Administration (NASA) and the Japan Aerospace Exploration Agency. Its satellite measures rainfall from space. Another example is NASA's CALIPSO satellite. It is helping scientists understand how clouds and airborne particles affect Earth's climate.

Nearness to Water

Climate also depends on how close a region is to an ocean, lake, or river. The climate of Arizona is mostly dry because the state has very few large bodies of water. On the other hand, coastal states such as Florida tend to be humid. The ocean water evaporates constantly, keeping the air moist. Places that are close to water also have fewer temperature extremes.

Winds and Ocean Currents

The temperature differences between low and high latitudes cause the air around us to move. Warm air rises, while cold air sinks. This movement of air creates winds that travel the globe.

You learned on page 9 that the equator receives the most intense sunlight. Therefore, the atmosphere is strongly heated at the equator. Winds move this warm air from the equator toward the poles. They also move cold air from the poles toward the equator.

If Earth did not spin on its axis, these winds would blow straight north and south. Instead, they move diagonally across the globe. The reliable paths of these global winds give them the name **prevailing winds**.

Winds can also move water. The friction from wind blowing over water can drag the water in the same direction. The one-way flow of a gas or liquid is called a **current**. In other words, air currents create ocean currents when they come into contact with each other.

Just as the temperature of the air depends on latitude, so does the temperature of the ocean surface. The same temperature differences that cause winds also cause the ocean to flow. Ocean currents—like global winds—transport heat from low latitudes to high latitudes.

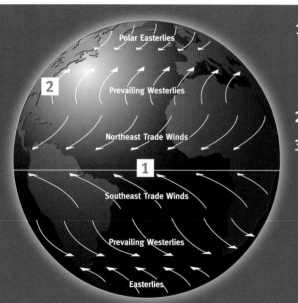

1. **Strong sunlight near the equator creates warm air currents that rise and flow toward the poles. At the poles, the air cools and sinks. The cooled air flows back toward the equator.**

2. **Earth's spin makes prevailing winds swerve east or west.**

3. **Some ocean currents move warm water from the equator to the poles. Others move cold water from the poles toward the equator. There are also currents that move along lines of latitude.**

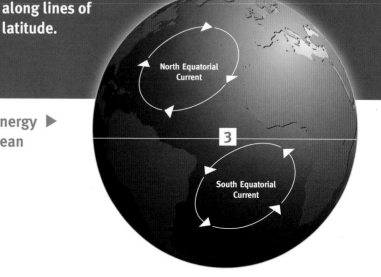

Winds and ocean currents transport heat energy ▶ between the equator and the poles. The ocean currents form circular patterns, known as gyres (JY-erz).

Summing Up

- Climate describes the average conditions of the atmosphere in a given region. To understand climate, observations of air temperature and precipitation must be made over many years.

- Weather, on the other hand, changes from day to day and even from moment to moment.

- The sun plays an extremely important role in Earth's climate, because its energy heats the atmosphere and gives rise to seasons.

- The greenhouse effect traps the sun's energy within the atmosphere and further warms Earth.

- The factors that affect air temperature and precipitation include latitude, elevation, topography, nearness to large bodies of water, global winds, and ocean currents.

Putting It All Together

Choose one of the activities below. Answer the question independently, in pairs, or in a small group. Share your responses with the class. Listen to other groups present their answers.

1 You read on page 9 that the equator receives the most heat energy from the sun. The sun never stops emitting energy. Why, then, don't areas near the equator continually increase in temperature? Draw a diagram that supports your explanation.

2 Reread the discussion of rain shadow on page 12. Suppose you are asked to model this effect for students in Grade 4. What materials might you use? Draw a diagram and list the steps that you would take to present your model to the class. List three follow-up questions that you could ask to check the students' understanding.

3 Using a globe or map of the world, find the approximate latitude of your city or town. Then locate a place on the same line of latitude, but in the opposite hemisphere. What conclusions can you draw about the climate of these two places? Use what you learned in Chapter 1 to support your conclusions.

Earth's Climate Zones

How are the major climate zones classified and described?

Would you like to visit Antarctica? Few people ever do. Antarctica is a continent that circles the South Pole. It is difficult to navigate through the surrounding ice-covered ocean. The weather is always cold and windy. Plus it is covered by snow and ice all year long.

Antarctica is in one of Earth's coldest **climate zones**. A climate zone is a region that has a specific range of temperature and precipitation throughout its extent. The simplest way to classify climate zones is to sort them into three major groups. These groups are polar, temperate, and tropical. The map on page 19 shows the major climate zones.

Scientists use a visual tool to display the characteristics of a climate zone. This tool is called a **climograph** (KLY-muh-graf). A climograph shows the monthly temperature and precipitation of a specific place. It usually combines a bar graph for precipitation with a line graph for temperature.

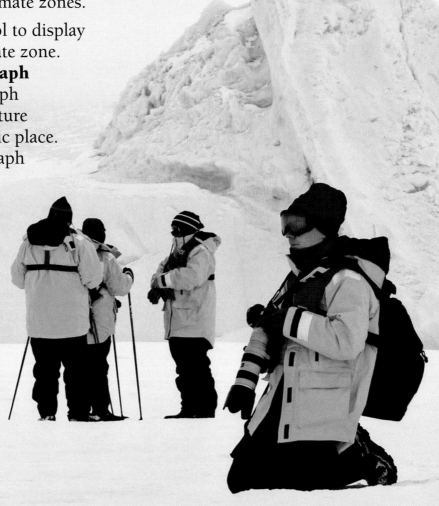

▲ The climate of Antarctica is too extreme for most forms of life.

The Root of the Meaning:

THE TERM

Temperate

The word **temperate** comes from the Latin word **temperatus**. This word means **"restrained"** or **"regulated."** It was originally used to describe people who exercised control over their behavior. The term was first used to describe climate in the early 1400s.

A climograph shows the annual climate of a specific place.

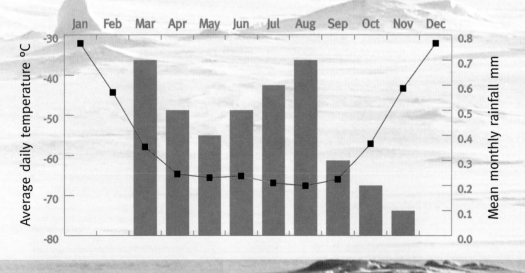

Average monthly temperature is plotted as a line, typically in units of °C.

Average monthly precipitation is usually in millimeters (mm) and drawn as a bar chart.

Antarctica is the most inhospitable place in the world. On this inland plateau, the air is too cold for precipitation to fall as rain. The amounts shown here are equivalent to how much water would fall if the snow were melted.

Three Major Climate Zones

The simplest way to classify climate is to divide the world into three zones, based on latitude. This system is useful for understanding the basic features of the world's climates.

Polar Zones

The **polar** zones surround Earth's North and South poles. The Arctic region is to the north, while the Antarctic region is to the south. Most polar climates have long, cold winters and short summers. Precipitation is low year-round.

Temperate Zones

Air masses from the surrounding two zones influence **temperate** climates. Tropical air masses move toward the poles, while polar air masses move toward the equator. Interactions between these air masses give temperate zones, or middle latitudes, moderate climates. Summers are warm, and winters are cool or cold.

Tropical Zones

Tropical zones—tropics, for short—lie within about 23.5° north and south of the equator. Here, temperatures are warm all year, and there is lots of rainfall.

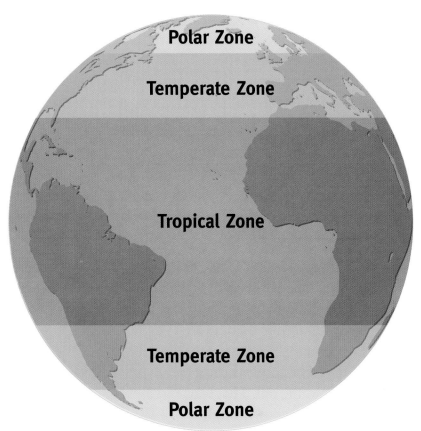

▲ Earth has one atmosphere with many climates. This map shows the three basic climate zones: polar, temperate, and tropical. Antarctica is in the southern polar zone.

✔ CHECKPOINT

Make Connections

In which climate zone do you live? Find out the average monthly temperature and precipitation for your town or city. Use the information to draw a climograph.

Life in a Climate Zone

Within each climate zone, you'll find distinct groups of plants and animals. That's because every living thing is adapted to the conditions in its environment. Only the hardiest plants can grow in the extreme cold and dryness of Antarctica. No land mammals can live in this polar zone, yet seventeen species of penguins call it home.

The tropics have more varieties of living things than any other zone. Think about the Amazon rain forest in equatorial South America. This warm region has high levels of rainfall and variable topography. It supports more kinds of plants and animals than any other place on Earth. A single bush may be home to more species of ants than are in the entire state of New Mexico!

Most places in temperate zones have four distinct seasons. This is true of much of the United States. Northeastern states may have Arctic-like winters and tropical summers. Many of the trees in temperate forests drop their leaves in the fall. This way, the trees can survive through the cold winter. When the temperatures warm in spring, the leaves begin to grow again. Flowers bloom and vegetables thrive during the hot, humid summers.

Science to Science: Earth Science and Biology

An important link exists between climate and the variety of life on Earth. Some plants grow only in warm, moist tropical climates. Others thrive in cool, temperate climates.

A small climate region with similar communities of plants and animals can form a biome (BY-ome). Life scientists recognize at least eight different biomes. These include grassland, temperate forest, tropical rain forest, tropical dry forest, savanna, tundra, taiga, and desert.

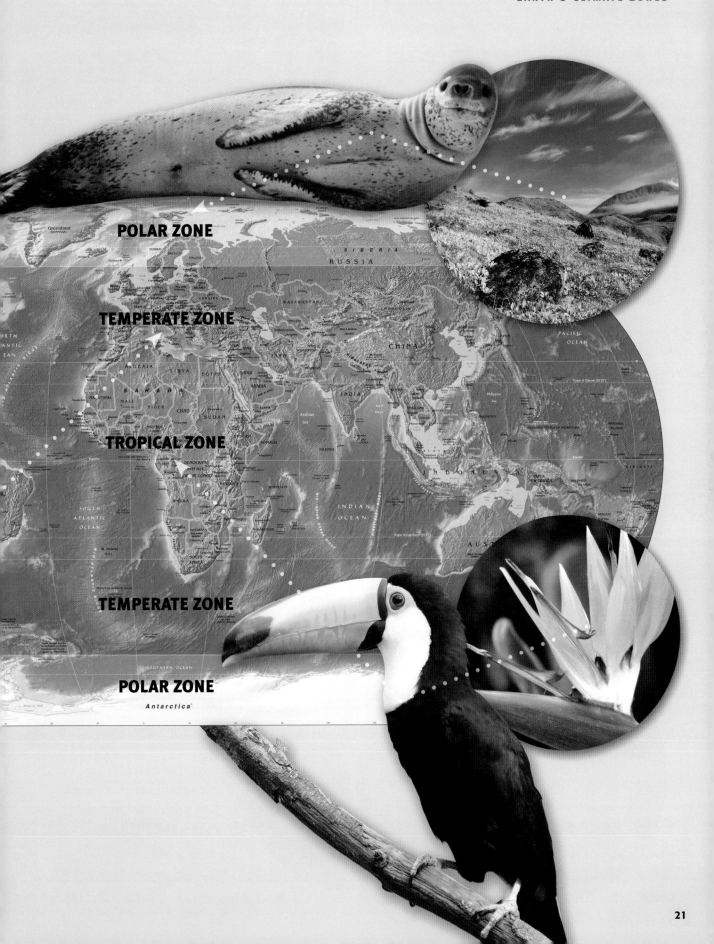

POLAR ZONE

TEMPERATE ZONE

TROPICAL ZONE

TEMPERATE ZONE

POLAR ZONE

Maritime and Continental Climates

Earth's oceans are very good at absorbing and storing the sun's heat energy. Oceans take in a lot of heat and hold it for a long time. However, the land is not nearly as effective. Land gets warmed by the sun only at the surface. Then it radiates the heat energy back out to the atmosphere quickly. It does not store the energy for long. This difference in heat absorption and storage sets up two very different climates depending on the nearness to water.

Coastal areas tend to have **maritime** climates. **Continental** climates usually occur inland. Water can store a lot of heat that it can release at a later time. Therefore, the ocean keeps maritime areas cool in summer and mild in winter. On the other hand, continental climates usually have hot summers and cold winters. The air also tends to be drier. Sometimes, if a coastal mountain range is present, a continental climate could prevail over a maritime climate. This effect is like the rain shadow that you read about on page 12.

They Made a Difference: Wladimir Köppen (1846–1940)

In 1918, the German scientist Wladimir Köppen was the first to publish a classification of climate zones. He divided the world's land areas into five major categories. These he based mostly on vegetation. He divided each category into several smaller categories to account for seasonal changes. Farmers found this system especially helpful.

You might be surprised to learn that Köppen was neither a geographer nor a climate scientist. He was a plant scientist. When Köppen was alive, few data were available for temperature and precipitation. However, the vegetation of the explored world was well understood. Köppen recognized that certain plants needed specific patterns of temperature and precipitation. He knew that mapping the distribution of plant types would, in the end, map climate as well. Today we have the kinds of detailed climate data that Köppen lacked. Even with this information, his original climate zones still hold up. Now over 100 years old, Köppen's classification scheme is still widely used.

◀ Because of its large mountain ranges, Alaska has both maritime and continental climates.

Hands-On Science:
Heating the Land and Water

The sun is the main source of heat energy on Earth. Does the sun's energy heat all of Earth's surfaces equally? Find out with this experiment.

TIME REQUIRED
20 minutes

GROUP SIZE
small groups

MATERIALS NEEDED

- 2 plastic containers
- 2 thermometers
- sand
- large tub of ice water
- tap water
- clock or stopwatch
- graph paper

PROCEDURE

1. Write a prediction about which holds heat longer— land or water.

2. Fill one plastic container with tap water at room temperature. Fill the other container to the same height with sand.

3. Place each thermometer about halfway down into the center of each container. Measure and record the initial temperatures.

4. Place both containers in a tub of ice water. Make sure the containers do not touch each other.

5. Wait two minutes, and then record the temperature in each container. Repeat this step every two minutes for ten minutes.

ANALYSIS

1. Graph your results to show the change in temperature over time in both containers. In which container did the temperature drop faster?

2. Compare your results to your prediction. Do they match? Write a brief report including the details of your investigation and what the evidence suggests about the heating of land and water. Explain the effect on climate.

MY OBSERVATIONS

Record your observations in a data table like this one.

Time (minutes)	Container 1 (water)	Container 2 (sand)
0		
2		
4		
6		
8		
10		

Subzones and Microclimates

Dividing the world into three zones is a very broad way to look at climate. Most places do not fit into such neat categories. For instance, deserts can be found in some tropical areas. In truth, each zone has places that combine characteristics of other climates.

Subzones are a way to organize the climate zones by dryness and temperature. Antarctica is a good example. It is classified as a polar zone, yet has three subzones. These are the interior region, the coastal areas, and the Antarctic Peninsula. The interior region has an extremely cold continental climate, with light snowfall. The coastal areas have somewhat milder temperatures and much more precipitation. The peninsula has a maritime climate.

A subzone may hold even smaller areas with climates quite unlike their surroundings. A **microclimate** is a miniature climate zone with its own unique characteristics. Microclimates can range in size from a protected garden to a city. They are affected mostly by large bodies of water and by topography.

For example, a concrete or tar-paved road in an urban area will absorb and reradiate a lot of heat energy. A tree-shaded park with grass and a pond located in the same urban area will be cooler and more moist than the surrounding city streets.

▲ A protected backyard garden or a city park can be a microclimate.

Summing Up

- A climate zone is a region that has a specific range of temperature and precipitation throughout its extent.

- A climograph shows the monthly climate conditions of a particular place.

- The simplest way to classify climate zones is by latitude.

- The three major zones are polar, temperate, and tropical. Because living things are adapted to the conditions of specific climates, each zone has a different collection of plants and animals.

- Climate zones can be further classified into subzones according to dryness and temperature. Maritime climates occur in coastal areas. Continental climates are found inland.

- Microclimates are small climate zones with characteristics that differ from the surrounding areas.

Putting It All Together

Choose one of the activities below. Respond to the prompt independently, in pairs, or in a small group. Share your responses with the class. Listen to other groups present their responses.

1 Choose one land area in a major climate zone. Find out what subzones occur within it. Draw a map and use different colors or patterns to show these subzones. Include a legend and brief description of each subzone so that anyone reading your map can understand your classifications.

2 Reread the discussion of maritime and continental climates on page 22. Research examples of these land climates. Choose one city or town that is representative of each. Using the library, newspapers, or a trusted Internet site, find out the monthly temperature and precipitation in both locations. Based on these data, draw a climograph for each location. Refer to the example on page 18 if you need to. Compare the two climographs in a paragraph.

3 You read about microclimates on page 24. Take a walk or bike ride around your school or neighborhood to search for a microclimate. Pay special attention to outdoor areas that are shaded, have a lower or higher elevation than the surrounding area, lie near a large body of water, or are sheltered from the wind. Draw a map of the area you observe, and indicate the characteristics of the microclimate. Explain how this area differs from the surrounding climate.

CLIMATE CONTROL

CARTOONIST'S NOTEBOOK • ILLUSTRATED BY PETE PACHOUMIS

... WE ARE BUILDING THE FIRST LUXURY UNDERWATER HOTEL, THE FIRST AIR-CONDITIONED BEACH ...

... AND THIS IS THE EARTH'S FIRST SEVEN-STAR, SAILBOAT-SHAPED HOTEL, YOU'LL NOTICE THE AIR-CONDITIONED SIDEWALKS, AND COOL-MIST FANS FOR YOUR COMFORT!

IN THE DISTANCE YOU WILL SEE OUR LATEST DEVELOPMENT — A MAN-MADE GROUP OF PRIVATE, AIR-CONDITIONED ISLANDS. WHEN COMPLETE, THIS MICROCLIMATE WILL RESEMBLE A MAP OF THE WORLD WHEN SEEN FROM ABOVE!

GOOD QUESTION!

HMM...

WOW! WHO PAYS THE ELECTRIC BILL?

Humans use energy to survive and also to be comfortable in different types of climates.

What comforts would you be willing to give up in order to conserve energy?

What comforts would you NOT be willing to sacrifice?

Changes in Climate

What are the causes and effects of climate change on Earth?

What might happen if short, mild winters replaced the long, snowy winters of New England? What if the yearly rainfall on the prairies of the Great Plains dried up, turning them into a desert? The changed climates would deeply affect the plants, animals, and people that live there.

If these descriptions sound like science fiction, think again. Scientists know that major changes in climate have occurred several times in Earth's past. Some living things were successful in adapting to the changes, but others were not. What causes Earth's climate to change? Do such changes happen all at once or slowly?

Natural Events That Cause Climate Change

What do earthquakes and ocean waves have in common? They are both natural events. A natural event results from the forces of nature. You can plan for it, but you cannot stop it. Certain natural events can cause changes in climate.

This map shows the giant continent of Pangaea that existed 225 million years ago. Over time, parts of Pangaea split off from one another, slowly forming the seven continents. ▶

PERMIAN
225 million years ago

TRIASSIC
200 million years ago

Continental Drift

Have you ever looked at a map of the world and noticed that the continents seem to fit together like a puzzle? That is just what the German explorer Alfred Wegener observed. In 1915, Wegener published his theory of **continental drift**. It explained how the continents formed. He suggested that the continents once belonged to a giant landmass that he named Pangaea (pan-JEE-uh). Pangaea is a Greek word meaning "all lands." Over millions of years, Wegener reasoned, the continents separated and drifted to their present locations.

JURASSIC
135 million years ago

CRETACEOUS
65 million years ago

PRESENT DAY

Wegener used **fossil** evidence to support his theory. A fossil is the remains of a living thing from long ago. He matched fossils on the edges of continents with those on the edges of facing continents. For instance, fossils from the reptile Mesosaurus occur in both Africa and South America. Wegener hypothesized that the only way Mesosaurus could have lived on both continents is if those continents were once connected.

Today we know that Wegener was correct. Scientific evidence shows that the continents are constantly on the move.

✔ CHECKPOINT
Read More About It
Scientists rely on fossil evidence to learn about Earth's past. Visit the library or use the Internet to find out how fossils form. Learn more about what fossils can tell about climate change. Present your findings in a short report.

The continents have indeed changed—and continue to change—their positions around the globe. These changes have affected Earth's climate.

CYNOGNATHUS FOUND IN SOUTH AMERICA AND AFRICA

GLOSSOPTERI FOUND IN AFRICA, ANTARCTICA, AND INDIA

LYSTROSAURUS FOUND IN SOUTHERN CONTINENTS

MESOSAURUS FOUND IN AFRICA AND SOUTH AMERICA

▲ Scientists use many kinds of evidence to support their theories. Wegener found similar fossils of plants and animals on different continents. He hypothesized that these continents were once connected.

Ice Ages

Nestled in the southern Black Hills of South Dakota is Wind Cave National Park. The park has relatively mild winters and warm summers with little rainfall. Animals such as bison and elk live in the park's prairies and pine forests. Why, then, are fossils of tropical ocean animals such as corals found here?

The answer to this mystery is twofold. Wind Cave was once located near the equator and was covered by shallow seas. Over millions of years, the region drifted to its present latitude of 43.5°N. But continental drift alone cannot account for the disappearance of the region's ocean. Falling sea levels can.

During Earth's history, the climate has varied between warm periods and cold periods. During extremely cold periods, or **ice ages**, there were large differences in temperature between the equator and the poles. Immense glaciers the size of continents covered portions of Earth. A glacier is a large mass of ice that moves very slowly.

These animals lived about 350 million years ago in tropical oceans. Their remains are found in the Black Hills of South Dakota at an elevation of 1,085 meters (3,560 feet).

colonial coral fossil

lampshell fossil

When Earth enters a warm period, the glaciers melt, at least partially. Water from the melting glaciers raises the sea level. Then, when Earth enters another cool period, much of its fresh water becomes frozen again in glaciers. Sea levels fall.

Scientists agree that there have been at least four major ice ages in Earth's past. Fossil plants and animals have recorded each change. Within several thousands of years, lush forests of tropical ferns changed to pine trees. Warm, shallow seas became cool mountains and caves.

Earth's last ice age began about one million years ago. The Great Ice Age, as it is called, reached its peak about 20,000 years ago. The last major ice sheet to spread across the United States melted about 6,000 years ago. Mountain glaciers are its last remains. Most of our environment, culture, and human history have been shaped by the events of this Great Ice Age.

Career: Climate Scientist

Climate scientists work all over the world, from the warm waters of the Caribbean Sea to the snowy lands of the South Pole. Some use satellites to observe and record global climate conditions. Others use climate data to make predictions. Climate scientists help governments and businesses make decisions that affect people and the environment. To work as a climate scientist, you need a college degree in math, science, economics, law, or engineering. Most climate scientists also earn an advanced degree after college.

▲ Many of Earth's land features, such as Yosemite Valley in California, were carved by glaciers.

El Niño

In Spanish, el niño means "the child." When both words are capitalized, **El Niño** refers to an atmospheric event that takes place along the coast of South America. During El Niño, a major change in the atmosphere reverses the direction of the prevailing winds.

In normal years, easterly trade winds blow above and below the equator. These winds blow from east to west, causing the ocean water below them to flow in the same direction. The currents move water away from the western part of South America.

This results in upwelling along the coastline. Upwelling is the upward movement of cold water from below the ocean surface. Upwelled water is rich in nutrients, which many food chains depend on. Every few years, however, this normal pattern reverses. When the winds change, the ocean flows in the opposite direction—toward the coast instead of away from it. Upwelling stops.

El Niño usually lasts between one and two years. No one knows exactly what causes El Niño, but one thing is certain: Although it begins at the equator, El Niño affects the world. It can damage economies and destroy environments. Severe storms and flooding may hit some areas, while other regions may get little or no rain at all for long periods. Without upwelling, the food chains that depend on the nutrient-rich water collapse. This can affect major fisheries on which the world depends.

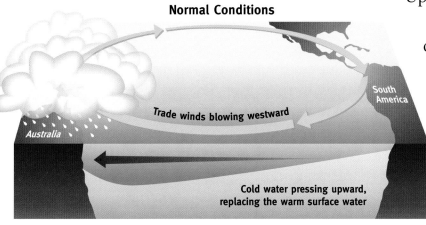

Normal Conditions

Trade winds blowing westward

Australia

South America

Cold water pressing upward, replacing the warm surface water

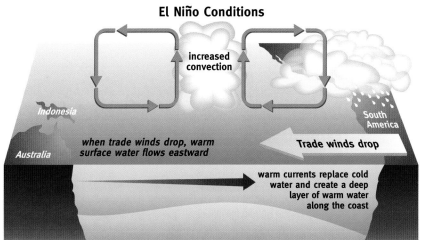

El Niño Conditions

increased convection

Indonesia

South America

Trade winds drop

when trade winds drop, warm surface water flows eastward

Australia

warm currents replace cold water and create a deep layer of warm water along the coast

▲ El Niño is a change in the atmosphere that starts along the equator but affects the entire world.

Science and Technology: Eye on El Niño

Scientists use different kinds of technology to predict and monitor El Niño. Some instruments are placed on floating devices called buoys (BOO-eez). There is a network of buoys in the tropical Pacific Ocean. These buoys collect data on wind direction and speed, air temperature, ocean temperature, currents, and rainfall. Satellites transmit the data in near real-time to science stations on land. Scientists analyze the data to make predictions and forecast the effects of El Niño so that people can prepare for it.

▲ This buoy is anchored in the ocean. It measures the direction and speed of the wind, as well as air and ocean temperatures.

Human Activities That Cause Climate Change

Think back to the greenhouse effect, which you read about in Chapter 1. Normally, this effect is helpful. After all, it keeps Earth from being a lifeless ball of rock. Without the greenhouse effect, the air temperature at Earth's surface would likely be about 33°C (60°F) colder than its present temperature. Life as we know it would be impossible.

What happens when there are too many greenhouse gases? The atmospheric "blanket" gets thicker. Earth gets warmer and warmer, at a faster rate than normal—a condition known as **global warming**.

Since the second Industrial Revolution of the late 19th century, greenhouse gases have been on the rise. Many scientists agree that the main cause of increased warming is an increase in carbon dioxide gas, or CO_2. This gas enters the atmosphere naturally from volcanic eruptions, forest fires, and respiration by plants and animals. But CO_2 also comes from some unnatural sources, which are problematic.

Greenhouse ▶ gases act like a blanket around Earth, trapping heat inside the atmosphere. The thicker the blanket, the warmer Earth gets.

Using Fossil Fuels

The inventions of the Industrial Revolution—like the steam engine, and later, the automobile—needed fuel to run. Most of the fuel (coal, oil, and natural gas) came from the fossils of plants and animals. Not surprisingly, they are referred to today as fossil fuels.

To get the energy stored in fossil fuels, the fuels need to be burned. Today we burn fossil fuels for heat, transportation, and electricity. This activity releases lots of greenhouse gases—mainly CO_2. Since the start of Industrial Revolution, the levels of greenhouse gases in the atmosphere have gone up each year. They are now about 25 percent (one-fourth) higher than what they were 150 years ago. In just the past 20 years, about 75 percent (three-fourths) of greenhouse gas releases have come from the burning of fossil fuels.

Some studies show that the rise in greenhouse gases has increased Earth's average temperature by at least 0.6°C (1.1°F). This might seem like a small amount. However, most scientists agree that this rise could be enough to contribute to global warming.

▼ This power plant is near the Grand Canyon in Arizona. It burns coal in order to provide electricity to nearby states.

Pollution

Pollution is the release of something that can damage Earth's resources. Releasing greenhouse gases is a form of air pollution. Other forms of air pollution can also harm the environment. Some buildings and factories have smokestacks that send chemicals and particles into the air. The particles can lower the amount of sunlight that enters and leaves the atmosphere. Other forms of air pollution can affect the clouds and precipitation in an area.

✔ CHECKPOINT

Make Connections

Think about how your life would be different if you had lived before the Industrial Revolution. What activities do you enjoy today that you could not have done in the mid-1700s and early 1800s? List the things or activities that you would miss the most.

▲ Smog is a form of air pollution. The term is a mix of the words "smoke" and "fog."

Everyday Science: What Are the Alternatives?

The burning of fossil fuels produces CO_2, which speeds up global warming. To reduce global warming, many people are using other energy sources—ones that produce little or no CO_2. These types of sources are called alternative energies. You may already be familiar with some of them.

- Solar energy (energy from the sun)

- Nuclear energy (from the splitting of atoms)

- Wind power

- Geothermal energy (from heat produced deep below Earth's surface)

- Biomass fuels (from plants and animal wastes)

- Water power (from the movement of flowing water)

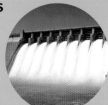

When a forest is cut down, the climate can change.

Deforestation

Think about how people use the land. They grow crops, herd animals, build cities, and pave highways. Some land uses contribute to climate change.

Deforestation is the cutting down of trees to clear an area. If the trees are not replanted, or if other forest areas are not set aside, then this practice can be harmful. For one thing, trees—like all plants—store carbon. When forests are cleared, and the trees either rot or are burned, this carbon is released as CO_2. Scientists estimate that deforestation makes up at least 20 percent (one-fifth) of all CO_2 released from human activities.

Trees also draw water up from the ground through their roots and release it into the atmosphere. Up to 30 percent (about one-third) of the rain that falls in a tropical rain forest is water that the forest has recycled into the atmosphere. When part of a forest is removed, the amount of precipitation in that region can change. As you have learned, precipitation affects a region's climate.

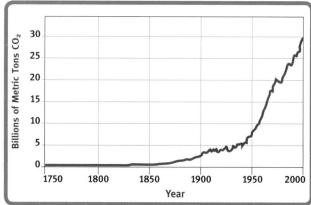

▲ Human activities add billions of metric tons of carbon to the atmosphere every year. This amount has increased steadily over the last 150 years.

Urbanization

If you live in a city, then you reside in an urban area. The homes and buildings in urban areas are built close together, with little open space. In an urban area, the human population is too dense to grow its own food.

A suburban area is a place just outside of a city. There are fewer buildings and more parks and recreational areas. People who live in suburban areas, or suburbs, often travel to the city for work.

The opposite of an urban area is a rural area. In a rural area, the homes and buildings are spread out. There is plenty of open space for land uses such as farming and ranching.

Urbanization is the conversion of rural or undeveloped lands into urban areas. Urbanization can create heat islands. These areas are warmer than nearby rural areas. The air temperature of a city with one million people can be 1 to 3°C (1.8 to 5.4°F) warmer than its surroundings. As you might expect, heat islands add to climate change.

▼ As the world's population grows, more and more heat islands form.

✔ CHECKPOINT

Think About It

Keep a journal for one day, listing all of your activities. How might those activities contribute to climate change? How might they help prevent climate change?

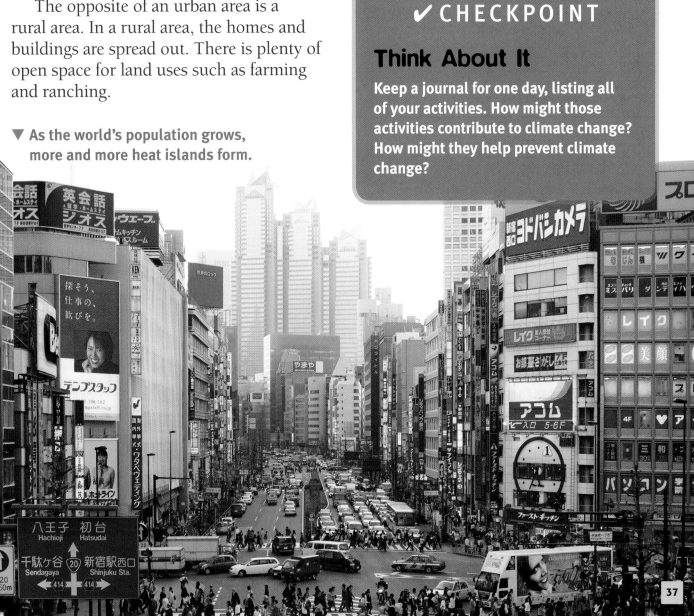

Possible Effects of Climate Change

What happens as Earth gets warmer? Scientists don't know exactly, but they can make educated guesses.

Melting Ice

In winter, a sheet of floating ice covers most of the Arctic Ocean. This sheet is larger than the United States and up to 3 meters (10 feet) thick. Some of the ice melts naturally in summer. However, more is melting now than ever before. The Arctic's winter ice sheet is shrinking by 9 percent every decade. In 2007, the sheet's area reached a record low. If this rate continues for a few more decades, the Arctic ice sheet could disappear entirely.

Melting sea ice does not raise the surrounding water level, because the ice displaces the exact amount of water it holds. However, it could cause faster melting. Ice acts like a mirror, reflecting the sun's rays. When sea ice melts, it gives way to darker water. The water absorbs sunlight and becomes warm, melting even more ice.

▼ The Greenland ice sheet covers about 1.7 million square kilometers (656,000 square miles). Scientists are less sure about the fate of the Greenland and Antarctic ice sheets than they are about Arctic sea ice.

A massive ice sheet covers the Antarctic continent, with large shelves extending over the ocean. Some of these shelves have been collapsing in recent years. The land ice, however, seems to be stable for now. A large ice sheet also covers Greenland. Scientists are studying both areas to learn more about how they are changing—and why.

Rising Sea Levels

As ice over the land melts, the water enters the oceans. The added water could cause sea levels to rise. Most scientists think the average rise will be about 48 centimeters (19 inches) by 2100. However, if the ice sheets of Antarctica or Greenland collapse entirely, sea levels could rise much higher. This will be a problem for people living in coastal areas around the world. Even a small rise in sea level can cause havoc for low-lying communities, especially in island nations.

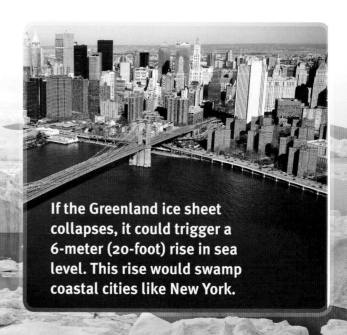

If the Greenland ice sheet collapses, it could trigger a 6-meter (20-foot) rise in sea level. This rise would swamp coastal cities like New York.

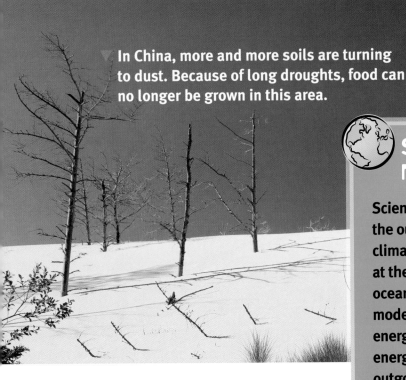

In China, more and more soils are turning to dust. Because of long droughts, food can no longer be grown in this area.

Science and Math: Making Predictions

Scientists use climate models to predict the outcomes of climate change. A climate model uses mathematics to look at the interactions of the atmosphere, oceans, land surface, and ice. All climate models take into account the incoming energy from the sun and the outgoing energy from Earth. If the incoming and outgoing energy are not balanced, then temperature changes will result. Some climate models are simple. Others are more complex. The most popular models today are the ones that simulate releases of carbon dioxide gas.

Extreme Weather

"Is it hot enough for you?" is one question you might be hearing more often. Extended periods of high temperatures, or heat waves, are becoming more common. In 2009, a two-week heat wave hit southeastern Australia, with record-breaking temperatures above 46°C (114°F). Prolonged heat waves can lead to wildfires, droughts (DROWTS), and famine. Droughts are long periods with little or no rainfall. They can gradually turn soil into dust.

Even a small rise in ocean temperatures can change the weather. Typhoons and hurricanes can become more intense, with higher wind speeds and heavier rainfall.

▲ The peacock butterfly is common in Scotland. Since 2000, it has extended its range farther and farther north.

Adapting to Changing Conditions

Climate change can upset the balance of nature. If living things do not adapt to changing conditions, they may die off. Scientists are already observing the loss of algae and resulting bleaching of corals as oceans get warmer. Rare butterflies, sensitive to rainfall, are vanishing in England.

As temperatures rise, more animals are expected to move northward to cooler climates. Trees may extend farther up mountains. However, if they cannot find food, water, or the shelter they need, Earth's living things will not survive the changing climate.

Hands-On Science: Melting Sea Ice

One possible consequence of global warming is the melting of sea ice. What happens to the atmosphere when the sea ice is gone? Find out with this experiment.

TIME REQUIRED
2–3 HOURS

GROUP SIZE
individual or small groups

MATERIALS NEEDED

- **2 thermal cups, such as foam cups or coffee mugs**
- **insulating lid for one of the cups**
- **2 thermometers**
- **boiling or near-boiling water**
- **clock or stopwatch**
- **graph paper**

▲ Sea ice over the Arctic Ocean next to Alaska

SAFETY CONSIDERATIONS

Be careful not to burn yourself with the hot water. Wear safety goggles when using liquids. Make sure your thermometers are suitable for measuring high temperatures so they do not break.

PROCEDURE

1. For one of the cups, make a lid that can fit around a thermometer. Use material that provides insulation, such as foam.

2. Fill the two thermal cups with boiling or nearly boiling water so they are each about 3/4 full. The water in these cups represents the polar oceans.

3. Place a thermometer in each cup. Place the lid around the thermometer on one of the cups. This lid represents sea ice covering the ocean.

4. Measure the temperature in each cup every 15 minutes for one to two hours. Record the data in a table like the one below.

ANALYSIS

1. Graph your results to show the change in temperature over time in both containers.

2. What did you observe to happen in each cup? Why did this happen?

3. When heat was transferred out of the "ocean," where did it go?

4. In what ways does this activity model how sea ice helps cool the air?

Record your observations in a data table like the one below.

MY OBSERVATIONS

Time	Ocean with sea ice (cup with lid)	Ocean without sea ice (cup without lid)
9:00 A.M.		
9:15 A.M.		
9:30 A.M.		
9:45 A.M.		
10:00 A.M.		
10:15 A.M.		

Summing Up

- Earth's climate is always changing. In the past, the climate has been warmer or cooler than today.

- When Earth was warmer, the oceans rose and covered much of the planet.

- Fossils show that most of these changes occurred slowly over hundreds of thousands of years.

- The natural events that cause climate change include continental drift, ice ages, and El Niño.

- Humans can also cause climate change by burning fossil fuels, pollution, deforestation, and urbanization.

- These activities release greenhouse gases into the atmosphere, which add to global warming.

- The possible effects of global warming include melting sea ice and glaciers, rising sea levels, more severe weather, and the loss of living things.

Putting It All Together

Choose one of the activities below. Respond to the prompt independently, in pairs, or in a small group. Share your responses with the class. Listen to other groups present their responses.

1 Think about the weather in your state or region over the past year. Make a judgment as to whether the previous season was warmer or colder than usual and if it was drier or wetter than normal. Consider the factors that might influence your judgment, such as how much time you spend outside. Then find the actual data for your location at the National Climatic Data Center or other trustworthy climate resource. Write a short essay comparing your statement to the data. In your essay, comment on how reliable human memory is compared to scientific measurements of temperature and precipitation.

2 Review the activities on pages 34–37 that cause global warming. What changes can you make in your daily life that might help prevent global warming? Make a poster that lists these actions. Decorate your poster with drawings and pictures. Display your poster so others can follow the steps you suggest.

3 On page 39, you read that living things must adapt to climate change or die. Choose a specific animal that is native to your region. Find out about the natural history of your animal. Where does it live? What does it eat? What eats it? What temperatures can it tolerate? How much water does it need? What other animals compete with it for food and living space? Then consider how climate change might affect your animal. Make a table listing changes such as higher temperatures, more heat waves, heavier rainfall or snowfall, flooding, drought, and sea-level rise. Predict how much each change might affect your animal and how your animal might adapt (such as moving north, laying eggs earlier, or using fewer breeding places).

Conclusion

▲ As climates change, these polar bears could face extinction.

What's happening in the Arctic today is beyond what any climate model has predicted. The thickness and extent of the ice sheet has changed radically over the past few decades. For the first time in recent history, the North Pole is an island. Enough ice has melted to allow ships to circumnavigate the Arctic ice cap.

Global warming is a serious problem for the animals that feed and breed on the sea ice of the Arctic. The most vulnerable of these is the polar bear. Polar bears depend on a chain of organisms—including seals, fish, shrimp, and algae—to live. What will be the polar bears' fate if this food chain disappears?

Science, as a rule, is never certain. Scientists strive to figure out what is probable, what is not probable, and what is doubtful. In the case of climate change, we know that Earth was warmer in the past. We do not know how much warmer Earth will get in the future. If global warming continues at its present rate, the consequences should be of concern to all.

How to Write a Scientific Explanation

Often, a scientist is asked to explain in writing how or why something happens the way it does. Why is the sky blue? What makes tree leaves change colors each fall? This type of writing is called a scientific explanation.

Many of the magazine articles in popular science magazines are scientific explanations. They are also found at museums and on some government Web sites. A good scientific explanation has the following elements:

> A title that states the problem or question

> An interesting opening

> An example with specific details

> Facts that support the answer

> Well-written sentences organized into paragraphs

> Diagrams or illustrations, if needed, with captions

> A concluding sentence or paragraph

If possible, the facts and details in a scientific explanation should come from a primary source. A primary source is a document or other form of information prepared by someone with direct knowledge of the subject.

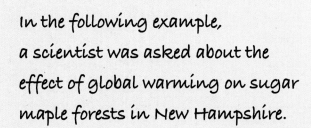

In the following example, a scientist was asked about the effect of global warming on sugar maple forests in New Hampshire.

HOW WILL GLOBAL WARMING AFFECT NORTHERN FORESTS?

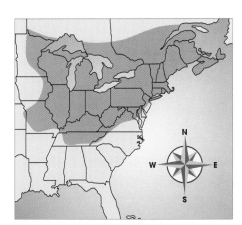

▲ the native range of sugar maple trees

Each fall, visitors flock to the White Mountains of New Hampshire to see the changing colors of the sugar maple trees. Some people are concerned that global warming might cause the sugar maples to migrate northward.

Sugar maples can only grow in regions with cool, moist climates. In the White Mountains, they reach elevations of about 760 meters (2,500 feet). The trees can tolerate temperatures from −40°C (−40°F) to 38° C (100°F). Occasionally, they can stand extremes that are 11°C (52°F) above or below this temperature range. They need about 510 to 2030 millimeters (20 to 80 inches) of precipitation each year.

Climate and tree-growth models suggest that a warmer climate may shrink the presence of sugar maples in their southern extent. However, the sugar maples of New Hampshire will likely take root in northern climates that meet the conditions stated above. Northern Canada could see thick forests of sugar maples if warming trends continue.

Glossary

climate	(KLY-mut) *noun* the average set of atmospheric conditions in a region over a long period of time (page 6)
climate zone	(KLY-mut ZONE) *noun* a region that has a similar range of temperature and precipitation throughout its extent (page 17)
climograph	(KLY-muh-graf) *noun* a graph that shows the monthly temperature and precipitation in a specific place (page 17)
continental	(kahn-tih-NEN-tul) *adjective* referring to inland areas (page 22)
continental drift	(kahn-tih-NEN-tul DRIFT) *noun* a theory, put forth by the German explorer Alfred Wegener, that explains how the continents formed by movements of Earth's crust (page 29)
current	(KER-ent) *noun* the directed flow of a liquid or gas (page 14)
El Niño	(EL NEE-nyoh) *noun* a major reversal of the prevailing trade winds that takes place in the atmosphere every few years, causing changes in climate (page 33)
fossil	(FAH-sul) *noun* the remains of a living thing from long ago (page 30)
global warming	(GLOH-bul WORM-ing) *noun* the steady warming of Earth caused by an increase in greenhouse gases in the atmosphere (page 34)

greenhouse effect	**(GREEN-hows ih-FEKT)** *noun* **the trapping of heat energy caused by certain substances in the atmosphere that keep Earth warm (page 10)**
ice age	**(ISE AJE)** *noun* **an extremely cold period in Earth's history (page 31)**
latitude	**(LA-tih-tood)** *noun* **a measure of distance from the equator, given in degrees (page 11)**
maritime	**(MAIR-ih-time)** *adjective* **referring to coastal areas (page 22)**
microclimate	**(MY-kroh-kly-mut)** *noun* **a miniature climate zone having its own unique climate characteristics (page 24)**
polar	**(POH-ler)** *adjective* **referring to the cool areas surrounding Earth's North and South poles (page 19)**
prevailing winds	**(prih-VAY-ling WINDZ)** *noun* **the known paths of the major winds that travel the globe (page 14)**
rain shadow	**(RANE SHA-doh)** *noun* **the effect of the nearness of the ocean and the height of a mountain, which creates a warm dry region inland of the mountain (page 12)**
temperate	**(TEM-puh-rut)** *adjective* **referring to the areas between Earth's polar and tropical zones (page 19)**
topography	**(tuh-PAH-gruh-fee)** *noun* **Earth's surface features, such as mountains or valleys (page 12)**
tropical	**(TRAH-pih-kul)** *adjective* **referring to the warm areas just north and south of the equator (page 19)**

Index